MW00962460

CURIOUS
REPTILES
and
AMPHIBIANS

The **Good** AND THE **Beautiful**

Written by Nyree Bevan | Designed by Phillip Colhouer
Cover design by Robin Fight

©2021 JENNY PHILLIPS | goodandbeautiful.com

Of all God's creatures, reptiles and amphibians are some of the most fascinating. While it's easy to simply refer to them as "lizards" or "snakes," there is so much more to reptiles than a couple types of animals. And though they are often confused with reptiles, amphibians have their own differentiating characteristics. From big to small and aggressive to meek, there are many unique and wonderful reptiles and amphibians. Let's take a look at some of them!

ALLIGATOR SNAPPING TURTLE

Alligator snapping turtles can live 50–100 years in the wild.

Because of its wild spiked shell, this turtle has been called the "dinosaur of the turtle world."

Most alligator snapping turtles spend all their lives in the water, except when the females go on land to nest.

They can hold their breath underwater for 40–50 minutes!

Alligator snapping turtles have tongue lures, bright red parts of their tongues that they extend and wriggle slightly. These look like worms, so fish and other prey come to eat them but are snatched up by the alligator snapping turtle instead!

This species of turtle is listed as threatened due to habitat loss and humans hunting to trade turtle shells.

TOADHEAD
AGAMA

Toadhead agamas have milky-white underbellies and black tips on their tails.

When a toadhead agama feels threatened, it can open the **lappets** on the sides of its head to make it look larger than it really is.

These agamas become dormant during the cold season and hibernate by digging single burrows.

Toadhead agamas are **insectivores**, eating many ants, flies, and locusts, as well as spiders.

PHILIPPINE SAILFIN LIZARD

Sailfin lizards are **oviparous**, which means they lay eggs, and often they lay their eggs in riverbanks.

Sailfin lizards have **pineal eyes**, also called third eyes, on the tops of their heads. Their purpose is still not known, but they are thought to be used as homing mechanisms.

The Philippine sailfin lizard is the largest lizard of the agama family and can grow up to 91 centimeters (36 inches) long.

These lizards have flat toes with scale fringes that allow them to run across water!

Their scientific name is **Hydrosaurus pustulatus**. *Hydro* means "water" and *saurus* means "lizard." These lizards often retreat from predators by dropping into water and swimming away.

CARPET PYTHON

Carpet pythons have a unique design of splotches and crossbands. This pattern (for which these snakes are named) often resembles an Oriental carpet.

There are many subspecies of carpet python, including the diamond python and the jungle carpet python.

Carpet pythons are found throughout Australia.

Carpet pythons have become popular pets.

A **clutch** of carpet pythons can have up to 40–50 eggs in it; they are kept warm by the mother "shivering" to generate heat within her muscles.

The biggest threat to carpet pythons is the cane toad, which contains a toxin that is fatal to them.

11

GHARIAL

With males reaching up to 6 meters (20 feet) long and weighing around 907 kilograms (2,000 pounds), gharials are one of the largest of the crocodilians.

Male gharials have large protrusions on the ends of their snouts, called **gharas** (Hindi for "mud pot"). Males use gharas to blow bubbles and vocalize in order to attract females during mating.

Gharials do not move very well on land because of their weak legs; they often move using what is called "belly-sliding," where they push their bodies across the ground.

Female gharials are the sole protectors of their eggs, and the hatchlings often stay with their mothers for weeks and sometimes even months.

Gharials are distinct from other crocodilians with their elongated snouts that have interlocking teeth on the upper and lower jaws. They use these sharp teeth to quickly catch fish and other prey!

PIG-NOSED TURTLE

Pig-nosed turtles are named for their large noses that resemble pigs' noses. This turtle's nose has receptors that help it locate prey and snorkel underwater.

Pig-nosed turtles are also known as fly river turtles.

Pig-nosed turtles are the only freshwater turtles that have webbed flippers like sea turtles.

Their shells have soft, leathery exteriors but are extremely hard underneath. This is a great protection from predators.

Female pig-nosed turtles often travel together when they are searching for a safe place to lay their eggs. (They can be very social turtles.)

Even if the eggs are ready and fully developed, they will not hatch until the rainy season, usually at midnight. This gives them the best chance for survival against predators.

BLUE CRESTED
LIZARD

The blue crested lizard is known for its beautiful turquoise-blue head and throat, but this color is only present during breeding season. At other times it is mostly reddish-brown or even grayish-brown.

Blue crested lizards are considered fully **arboreal**, which means they spend all their time in trees.

They are also **diurnal**—their activity time is during the day, and their sleeping time is at night.

Blue crested lizards are found in Southeast Asia.

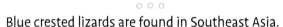

These lizards have long white stripes just under the eyes, which extend from their snouts to their shoulders.

TOKAY
GECKO

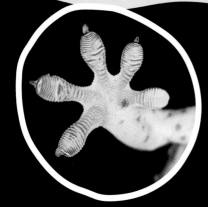

Because of many tiny hairs on each of their padded toes, tokay geckos' toes have amazing sticking power, capable of supporting up to 204 kilograms (450 pounds)!

Tokay geckos will lay their eggs on the sides of rocks or other similar surfaces. These sticky eggs harden as soon as they are laid, helping to protect the growing geckos inside.

They communicate using a range of noises from small chirps to loud whistles, and they also hiss or croak as a warning when feeling threatened.

The tokay gecko is known to bite when it feels threatened; watch out!

A tokay gecko can break off its tail at many segments as a means of defense. The cast-off tail continues to wiggle for several minutes, which distracts the predator and allows the gecko time to make its retreat.

These geckos are able to keep from casting shadows when on trees by opening their skin folds completely so they can blend in.

OLM
SALAMANDER

Olms are mostly blind because they live in caves or in underground streams. Their sight stops developing as they grow, and layers of skin even grow over their eyes.

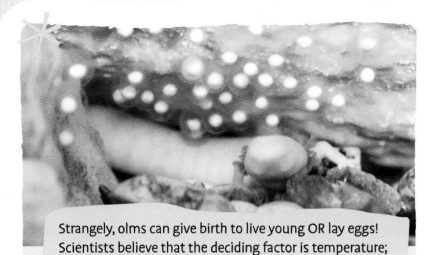

Strangely, olms can give birth to live young OR lay eggs! Scientists believe that the deciding factor is temperature; colder temperatures promote live births.

Using "super" senses such as highly developed smell and hearing, olms hunt and capture food. They may even be able to detect electric fields.

Because they are almost always without light, their skin loses most of its pigment, making them light pink, almost translucent, in color.

Olms have a very long life expectancy, averaging about 69 years, but it is believed that some have lived over 100 years!

STRAWBERRY
POISON DART
FROG

These frogs are also called "blue jeans frogs" because it looks like they are wearing a pair of denim jeans.

Hatched tadpoles are taken by the mother strawberry poison dart frog, one by one, from the leaf where they were laid to a pool formed in bromeliad leaves. There she lays unfertilized eggs in the water to feed them.

- Strawberry poison dart frogs have symmetrical patterns on their skin.

- Frogs of this species are quite small, growing to only around 3 centimeters (1 inch) in length.

- Males are very territorial, and if another male comes into their territory, they wrestle until one is pinned. The unsuccessful frog will surrender the territory to the dominant male.

GREEN TREE
PYTHON

When a green tree python is born, it is bright yellow or deep red. Its skin color doesn't change for 6–8 months, and then it turns green.

With **thermosensory pits** on their jaws, green tree pythons can track warm-blooded prey.

New Guinea

Australia

Green tree pythons are found in Australia and New Guinea.

24

Green tree pythons have been known to use strategies during hunting. One such strategy is sitting very still on a branch and dangling just the end of its tail. As animals approach curiously, the python strikes!

These snakes tend to rest wrapped around branches with their heads lying in the centers of their coils.

RED-EYED TREE FROG

Red-eyed tree frogs are very colorful with neon green bodies, bright red eyes, vivid orange toes, and beautiful blue and yellow coloring on their sides and upper legs. Alarming to predators, this coloration is part of their defense.

Female frogs lay their eggs in special places, usually on leaves over water. When hatching time comes, the tadpoles wiggle very quickly, which cracks open their eggs, and then they slide into the water to continue to grow and develop.

Red-eyed tree frogs are nocturnal and sleep with their eyes closed. They cling to the undersides of leaves to protect themselves by hiding.

The red color in their eyes comes from a third eyelid called a **nictitating membrane**, which covers their eyes for protection.

Their long, thin back legs are better for tree climbing than swimming, so most red-eyed tree frogs spend their adult lives in trees.

PANTHER CHAMELEON

Male panther chameleons sport brightly colored skins, while most females are light green, tan, or gray.

Panther chameleons have very large eyes that are covered with scaly skin with only a small opening to see. As with other chameleons, these eyes can move independently from one another.

There are several different species of panther chameleons, with many different colors depending on where they are from. They are some of the most colorful chameleons, displaying the many colors of the rainbow!

With tongues that can extend longer than their body length with sticky "suction cups" on the ends, panther chameleons can quickly grab prey such as insects or even small birds. They can extend and retract their tongues at fast speeds, much like a whip.

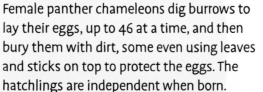

Female panther chameleons dig burrows to lay their eggs, up to 46 at a time, and then bury them with dirt, some even using leaves and sticks on top to protect the eggs. The hatchlings are independent when born.

A panther chameleon's skin has two layers of crystal-containing cells that can be stretched or relaxed, allowing the animal to quickly change its skin color by changing the reflected light.

RED-EYED CROCODILE SKINK

The red-eyed crocodile skink may look a lot like a crocodile, but it is in fact a skink. It only grows up to 25 centimeters (10 inches) long, and it doesn't have any sharp teeth.

However, like crocodiles, crocodile skinks have bony plates on their heads and spines down their backs, which act like armor and protect them.

Mother crocodile skinks are very protective! After laying only one egg at a time, the mother wraps herself around the egg and keeps it warm and safe. She will also take care of the hatchling for up to two weeks.

These lizards have been called "scaredy-lizards" because they seem to avoid fighting. If they feel threatened, they will "bark" to scare predators, but if that doesn't work, they might vomit and shed their tail. They've also been known to play dead to try to confuse predators.

Crocodile skinks seem to be quite shy and don't like being out of their hiding places.

FIRE SALAMANDER

Be careful around these critters; they can spray poison out of glands located behind their eyes. They can also release poisonous toxins through their skin!

Fire salamanders are members of the genus *Salamandra* and are known as lungless salamanders because they breathe not through lungs, but by absorbing oxygen through their skin.

These salamanders give birth to live larvae, one of the only species of salamanders to do so.

Fire salamanders are named after old legends. One legend said they were capable of withstanding fire; another said they were born in fire. Neither legend is true.

Fire salamanders eat several things, even their own skin after they shed it!

A fungus called Bsal, discovered in 2013, is one of the biggest threats to these salamanders. When infected they cannot eat or absorb oxygen through their skin.

GRAND CAYMAN
IGUANA

After breeding, a female will drive other iguanas—even much larger ones—out of her territory.

Growing up to around 2 meters (7 feet) long and weighing more than 11 kilograms (24 pounds), this iguana is the largest native land vertebrate of the Grand Cayman Islands.

The beautiful blue-green coloring of the Grand Cayman iguanas helps to **camouflage** them among the plants and rocks native to their homeland.

The red eyes of blue iguanas help to protect their pupils from bright sunlight.

These iguanas often use their strong, spiny tails as whips in defense against predators!

Grand Cayman iguanas are one of 11 species of "rock iguanas." They are diurnal (active during the day).

Grand Cayman iguanas are **pollinators**, eating many leafy plants and fruits and then dispersing the seeds.

BLUE VIPER SNAKE

Blue vipers are a species of pit vipers.

Viper snakes are often shorter and more sturdy than other snakes such as cobras. Vipers use this strength to ambush and attack their prey.

Pit vipers detect electromagnetic radiation through pit organs located on their upper lips. This helps them find and identify prey.

Instead of seeing light as humans do, pit vipers sense **infrared** light, most likely seeing body heat that helps them locate and judge the size of prey and predators, even in the dark.

Blue vipers are very rare, most often found in Indonesia. Green is a much more common color for pit vipers.

Living most of their lives in trees, blue pit vipers rarely come down except during mating season.

GLASS FROGS

Glass frogs are so named because of their translucent (or in some cases almost transparent) skin. In some species the internal organs can be seen, including their beating hearts.

After the eggs of glass frogs are laid (around 30 in one clutch), the father stays close by to protect them and keep them from drying out.

Glass frogs live mostly in rainforests, needing very moist living conditions, and often spend most of their adult lives high in the tree canopies.

There are more than 120 species of glass frogs, including emerald glass frogs, ghost glass frogs, and red-spotted glass frogs.

Due to their excellent eyesight, glass frogs are able to easily catch prey, such as insects and spiders. Because they help control the number of insects in the rainforest, they are a valuable part of the ecosystem.

VEILED
CHAMELEON

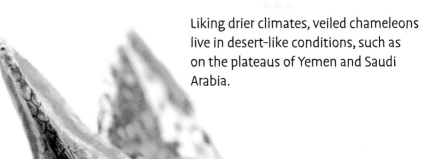

Liking drier climates, veiled chameleons live in desert-like conditions, such as on the plateaus of Yemen and Saudi Arabia.

Veiled chameleons are characterized by **casques**, or head crests, located on the tops of their heads. Although these casques help collect water droplets at night, their full purpose is still unknown.

When they feel threatened, veiled chameleons may curl up into tight balls and darken their color.

Unlike most chameleons, veiled chameleons enjoy plants as well as insects.

Veiled chameleons are one of the most aggressive chameleons, with males fighting for territories and mates.

RED-SPOTTED NEWT

Larval stage red-spotted newts live in water, receiving oxygen through feathery gills on their necks. When they metamorphose into juveniles, called red efts, they leave the water, and their gills are replaced with lungs. In one to seven years, they will return to the water but continue to breathe air through their lungs.

As red efts, this breed of newts is bright orangish with red spots outlined in black. This coloration warns predators. It also contains a neurotoxin that can be deadly to some animals and unpalatable to others.

Females lay at least 200 eggs, but they only lay a few eggs at a time, making this a long process.

The red-spotted newt is fairly small, and as an adult it only grows to 8–13 centimeters (3–5 inches) in length.

WEB-FOOTED GECKO

Living mostly in the Namib Desert, this sandy-colored, nearly translucent, web-footed gecko blends in perfectly.

To help keep their lidless eyes moist, web-footed geckos collect dew drops on their eyes and continually lick them with their long, light pink tongues to keep them clean.

○ ○ ○

Webbed feet help these geckos travel on top of sand. They also help them dig into the sand to bury themselves, which is where they spend most of the day to keep cool and sleep.

○ ○ ○

The web between their toes is fleshy but contains small cartilages that help coordinate the many muscles of their feet. This helps them to "scoop" the sand in order to bury themselves quickly.

○ ○ ○

With their large eyes and vertical pupils, these geckos can see very well at night when they are out hunting.

○ ○ ○

These adorable geckos only reach about 10–13 centimeters (4–5 inches) long.

BLACK-BREASTED LEAF TURTLE

Black-breasted leaf turtles have eyes that can move independently, just like chameleons.

Leaf turtles get their name from their flat shells with rough edges in the front and the back, resembling leaves.

The females only lay one to two eggs per clutch; these eggs have an elongated shape.

Black-breasted leaf turtles are among the smallest turtles in the world, only growing to about 13 centimeters (5 inches) long.

Male species of this turtle have white irises, whereas the females have tan irises.

47

GOLD DUST
DAY GECKO

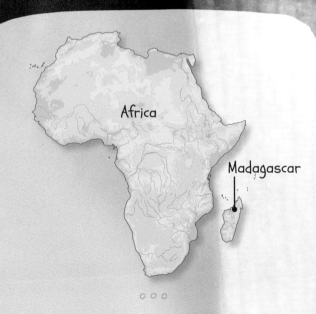

Africa

Madagascar

Day geckos are native to Madagascar and small islands off the coast of Africa.

Gold dust day geckos have a spattering of gold dots along their necks and shoulders over a beautiful green skin and have been called "living jewels."

These geckos not only eat insects like most geckos but also lick nectar from flowers and juices from overripe fruit.

The stunning sky-blue coloring above their big eyes looks almost like eye shadow.

Unlike most geckos, the day gecko is active during the day and sleeps at night.

FRILLED NECK LIZARD

When feeling threatened, frilled lizards expand the pleated skin flaps around their heads, rise up on their back legs, and hiss.

It is thought that their frills might also be used to regulate their body temperatures.

○ ○ ○

Another means of defense is their quick retreat, often with their frills extended and mouths open.

○ ○ ○

Interestingly, frilled lizards run upright on their hind legs. This is called **bipedalism**.

THORNY DEVIL

Conical spikes cover the thorny devil's entire body. These spikes are not bone, but are made of **keratin**, like your fingernails.

Using their sticky tongues, thorny devils almost solely eat ants, usually eating thousands in one day!

Each thorny devil has a "false head" on the top of its head. When it feels threatened, it puts its head between its front legs, curls up, and leaves this false head exposed in an attempt to confuse predators.

Thorny devils have a very slow, strange walk. They stop often and rock back and forth between steps.

Puffing up their chests, these lizards try to appear bigger than they are to ward off predators.

MADAGASCAR LEAF-NOSED SNAKE

These snakes are **endemic**, or native, to Madagascar.

Madagascar

With their flat, leaf-shaped noses and heads, they can easily blend into their environments, a great camouflage.

○ ○ ○
They are also known as Malagasy leaf-nosed snakes.

○ ○ ○
These snakes have rear fangs and are venomous, but their venom is not fatal to humans.

○ ○ ○
Leaf-nosed snakes hang upside down and either sway in the wind or hold still, resembling vines or twigs.

MATA MATA TURTLE

The **carapace**, or shell, sets this turtle apart with its big ridges and cone-shaped points. Some believe this is to resemble bark or rotting leaves, allowing it to blend into the river bottom.

They have webbed claws on each of their feet, the front feet having five claws and the back feet having only four.

Mata mata turtles rarely swim, preferring to walk underwater and rest among foliage to hide from prey. Their method of eating is called "vacuuming"; when prey comes close, they open their jaws and create suction, sucking the prey right in.

With snouts much like snorkels and very long necks, these turtles can stay underwater for a long time while continuing to breathe.

The mata mata turtle is a freshwater turtle found in the Amazon.

KNOB-TAILED GECKO

There are several species of knob-tailed geckos, including prickly knob-tailed geckos, smooth knob-tailed geckos, and centralian rough knob-tailed geckos.

Knob-tailed geckos are found throughout southern Australia due to its desert and sandy terrain.

When threatened, the knob-tailed gecko can bark very loudly, attempting to scare predators.

Their outer digits (or claws) are opposable, and their claws are non-retractable.

Using their sharp claws, knob-tailed geckos make sealable burrows to rest in during the hot day, coming out to hunt as night comes and the temperature drops.

They are named for the distinct knobs at the ends of their tails.